Pellegrino De Ros

The ideoplastic
evolution
of living species

Informative essay

This book is the translation of:
L'evoluzione ideoplastica delle specie viventi
ISBN 978-88-67512-91-1.
(**Pellegrino De Rosa**. Youcanprint editore, Italy. 2012).
All rights reserved to the author.
email: studio.derosa@alice.it
Address: Cav. Prof. Pellegrino De Rosa
Viale Michelangelo, 3
83020 - Sirignano (Av) - ITALY

Translated by **Antonio De Rosa**. *Sirignano (Av). Italy.(2012).*

ISBN | 978-88-91193-39-1

(Cover picture: *Periophthalmus Barbarus* or Mudskipper)

TAGS:

Evolution, Evolutionism, Intelligent Design, Darwin, Lamarck, Wallace, Bohm, Pribram, Aspect, mimicry, Giordano Bruno, monoideism, Evolutionary plasticity, Ideoplastic evolutionism, psychically-oriented evolution, self-induced evolution, intellectual coherence, plant neurobiology, SAM, quantum somatization, hypergenomic complessione, bio-entanglement.

I dedicate this book
to my son Antonio De Rosa
with imperishable love and affection.

The task of intellectuals is
than ever today, to sow doubt,
not to collect certainties.
(Noberto Bobbio)

Science helps us to recognize
the stamp of God.
(Pope Benedict XVI)

All truth passes through three stages.
First, it is ridiculed.
Then, it is violently opposed.
Finally, it is accepted as being self-evident.
(Arthur Schopenhauer)

Summary

Introductory pages

A *made in Italy* hypothesis on the evolution of living species called *Evolutionary plasticity* or *Ideoplastic evolutionism* has been presented in this informative essay as a possible third option between the opposing positions of neo-Darwinists and Intelligent Design supporters.

This new hypothesis of study was presented in 2009 by Dr. Pellegrino De Rosa, who is an italian agronomist, botanist, journalist and novelist.

It intends to answer some specific questions that arose from naturalistic observations, such as:

1. How the leaf-bug managed to make its body identical, even in the slightest detail, to the leaves of its natural habitat? Was it achieved only through random mutations and the consequent natural selection? Or was it achieved through the action of some unknown mechanism?

2. It is common knowledge that natural selection acts selectively on the distribution of mimetic individuals; but is there a more direct and not yet studied correlation between mimicry and the evolution of living species?

3. Apart from affecting genes expression, may the psyche of individuals also induce the appearance of entirely new genes?

As can be deduced both from this introductory essay and from the short story in the appendix, the author tried to answer these questions by adopting a multidisciplinary approach, that is to say by taking into account not only his personal naturalistic observations and the scientific data available, but also some philosophical, theological and physico-theoretical aspects.

Pellegrino De Rosa does not question at all the phenomenon of evolution, nor the results of genetics or the actions of some ecological mechanisms such as natural selection (however, he suggests reevaluating their importance in evolution). Yet, he believes that mutations in the genetic set of living species (and thus also their evolution) are caused neither by accidental or random variations nor by direct action of any God.

It is caused by *the ideoplastic action of the psyche of individuals on the genome and the epigenome.*

This hypothesis does not stem from abstract theoretical speculations; it is based on objective and repeatable observations of the behavior of rapid-mimetic organisms (such as the cuttlefish) and of the shape of the body of crypto-mimetic organisms (such as the leaf-bug). And only in a second phase it deals with these naturalistic observations establishing new links with other disciplines.

In short, according to the *Evolutionary Plasticity*, the force that causes the evolution is the same force that determines both temporary and permanent changes on the body of mimetic organisms. Therefore, it would be an alleged ideoplastic and mutagenic action of living beings' psyche. So, evolution would just be 'a will that takes shape'.

So, according to this hypothesis, the psyche of living beings would not only be able to regulate certain physiological functions and to epigenetically condition the expression of the existing genes (this has already been proved, for example, by the placebo effect and by the somatization phenomena observed in the hypnotic practice and in the MPD, or Multiple Personality Disorder), but also to create entirely new genes.

But this approach immediately poses certain problems.

The first problem: if the mind determines the evolutionary mutations, does it mean that all the organisms that have evolved, including plants, have a mind?

The second problem: What is the nature of the alleged interface between mind and genes?

The third one: when referring to the appearance of complex mutations that has no previous pattern in nature, what source of information the individual's mind access to?

The naturalistic observation of the existence of the amazing leaf-bug is an objective and indisputable fact that can be easily verified by anyone; but according to the author, it is necessary to establish a multidisciplinary and consistent connection with the so-called *frontier sciences* in order to answer the questions previously presented.

In order to understand the author's intellectual position you need to briefly dwell on his speculative leaning, which he called *intellectual coherence*.

In short, he states that if some findings are accepted in certain scientific and speculative circles, they should be equally valid in other circles.

Therefore, if the effect of the psyche on the body is accepted in hypnology and in psychiatry, this same effect must be also considered real when studying the evolution of living species.

And, if some concepts of cosmology and of quantum mechanics are considered acceptable when studying the inanimate matter (wave function collapse, interfering patterns, holographic paradigms, non-locality, etc..), they must be also considered valid when studying biological systems and evolution. Particularly, if quantum mechanics accepts that the observer can influence the dual behavior (undulatory and corpuscular) of the matter, it must also be accepted that a similar phenomenon can occur in biological systems. Therefore, it is not correct to hastily discard the hypothesis that the mind of the observer may also affect the biological matter, including DNA which has a quantum structure too.

Finally (and this applies only to believers), if you believe in a God, you must keep doing so even when you leave the place of worship and go to live your life and wear the white coat of scientist or researcher. According to the author, the person that does not do that, would suffer some kind of ideological schizophrenia.

Having said this, Pellegrino De Rosa responds to the first question by assuming that even plants have a mind, and strengthens his argument by referring to some researches on plant neurobiology carried out by Prof. Stefano Mancuso, of the University of Florence, and to some observations of Darwin himself.

Then, he refers to quantum biology and to a number of signs (neurological, physical, psychological and biological) mentioned in the text in order to answer the second question.

Finally, he refers to *Bohm's holographic paradigm* and to the philosophic thinking of Plato, Hegel and Giordano Bruno to respond to the third question.

Although he had a scientific background, he makes clear that does not believe science is the ultimate expression of human intellect, especially when it degenerates into scientism. He puts philosophy above it, and puts intuition and art even higher. He points out that all the great scientific geniuses of the past had their minds open to question, to the exploration of new possibilities and to philosophical thought. On the other hand, those he calls SAM (acronym for 'Middle Academic Scholars', that is to say those people who call themselves scientists, but often are just laboratory technicians or mere professors who have never discovered or proposed anything relevant) are ready to reject, sometimes even with arrogant rudeness, any new hypothesis of study proposed to them.

The author particularly denies the dogma of neo-Darwinist theories which holds that the blind chance is the source of evolution. He makes clear that all the observable phenomena in nature are regulated by laws that can be more or less complex. Then, he wonders why is the evolution of living species the only one caused by fortuity?

He describes such position as conformist, ideological and completely unscientific. He points out that the hypothesis of the appearance of complex and functional mutations caused by chance has never been scientifically proved neither in laboratory nor in ecological context, and supports it with examples. He emphasizes that considering random variations as the explanation of the evolution of complex organs is statistically unacceptable.

In fact, he admits that natural selection eliminates the 'less fit' and favors the survival and the spread of the 'fittest', but he makes clear that it only occurs if both competitors are present.

Therefore, natural selection must not be confused with the original cause of evolution, which is based on the force (which nature, according to the author, although still controversial, is psychic and undulatory) that led to the appearance of mutated individuals.

In order to be more specific, the author emphasizes the difference between the normal genetic variability observed within the same race and the appearance of complex evolutionary variations that lead to the formation of new species, and ends up accepting - in broad terms - Darwin's theory as the 'ecological theory of races', but rejecting it as the 'theory of evolution'.

Finally, the author accepts that genetics plays an extremely important role and states that he believes genes *are the vehicle and not the source of evolution*, and invites scientists to carry out experiments aimed at investigating whether the human mind, under special 'monoideistic' conditions or stress or altered consciousness, can induce functional mutations in the DNA, in the epigenome or in the hypergenome.

He also stresses that uncritically accepting Darwin's dogmas can seriously damage the scientific progress as it leads to ridicule and to discard a priori any other hypothesis of study and discourages the search for alternative explanations of the mysterious evolutionary phenomena.

And now Pellegrino De Rosa starts narrating, in an informal manner and narrative style, how his theory emerged and explains the main features, the relationships with other evolutionary theories and the links with related disciplines.

Since the aim of this essay is not to refute other people's theories but to present his hypothesis, he never dwells too much on the confutation of other theories, whether they confirm or oppose the evolution of living species.

Therefore, the few references he makes to other theories have the sole purpose of helping to clarify their differences and points of contact with the 'Evolutionary Plasticity' and have no claim to completeness. So, he refers those who want to deeply study the criticism of such theories (either creationist or evolutionary) that are not always shared, to the hundreds of thousands of texts published on these topics.

The appendix includes the most *FAQs* on 'Evolutionary Plasticity' taken from the responses provided by the author in several newspaper articles, in *social networks* and during the presentations of his books.

Other aspects of this theory are dealt with in his essays '*Evolutionary Plasticity. A new evolutionary hypothesis based on quantum biology and on holographic entanglement*' and '*What if Darwin was wrong?*'.

It is also explained in his novel '*Metamorfer. La gemma di Darwin*', which is set in the magical atmosphere of Naples and its beautiful Gulf. Despite being above all a fun and exciting fantasy-thriller that is full of humor and elegant eroticism, it leaks out a number of evolutionary and philosophical concepts that are not discussed in this essay.

(Enzo Pecorelli - journalist)

Bibliography:

P. De Rosa - *Leggendo una foglia.* (2009). L'Espresso publishing group. Italy.
(Reg. SIAE-OLAF: 2009001816 and 2009005213).
P. De Rosa - *Metamorfer. La gemma di Darwin.* (2011). Ed. Simple. Italy.
ISBN 978-88-6259-399-1.
P. De Rosa - *Plasticismo evolutivo. Una nuova ipotesi evoluzionistica basata sulla biologia quantistica e sull'entanglement olografico.* (2011). Ed. Simple. Italy.
ISBN 9788862594165.
P. De Rosa - *Metamorfer. La gemma di Darwin.* (2012). Ed. Youcanprint. Italy.
ISBN 9788866188650.
P. De Rosa - *E se Darwin si fosse sbagliato?* (2012). Ed. Youcanprint. Italy.
ISBN 9788866188704.

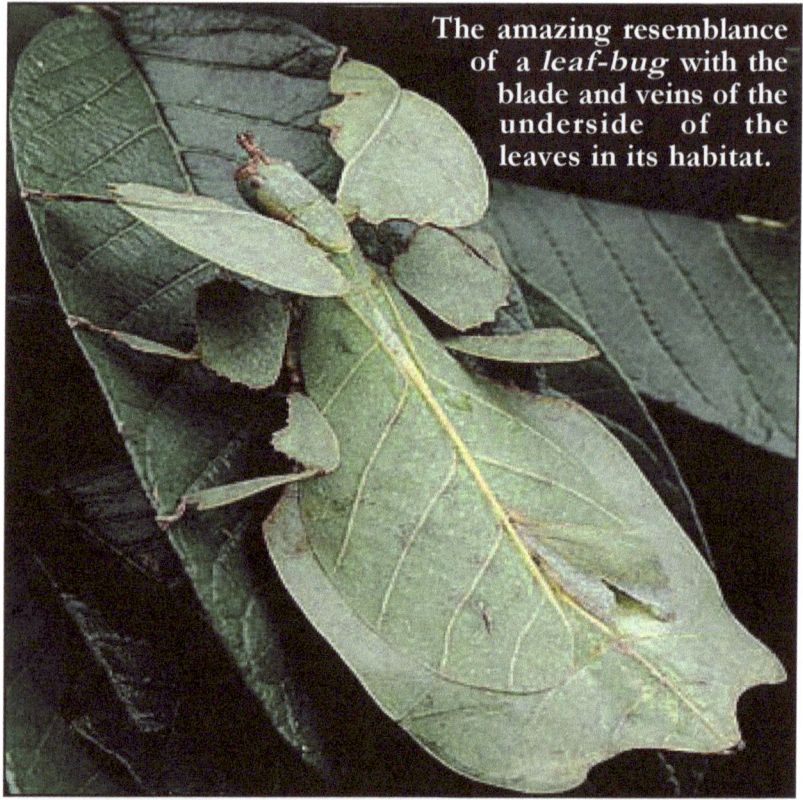

The amazing resemblance of a *leaf-bug* with the blade and veins of the underside of the leaves in its habitat.

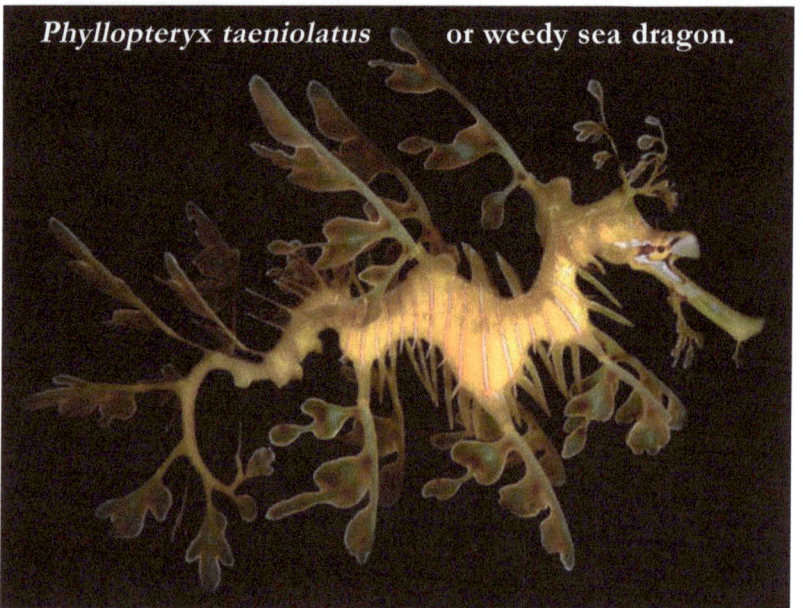

Phyllopteryx taeniolatus or weedy sea dragon.

Introduction

The fact that living species have not always been as we know them and have actually evolved with time, is considered objective and almost universally accepted, and I fully agree with it. The existence of fossils and some other undeniable facts such as the evolutionary convergence and the presence of vestigial genes - apart from all kinds of interpretation- prove it beyond any reasonable doubt.

However, it does not exclude at all the possibility that the evolutionary boost has an intelligent cause - not necessarily divine - and that is not caused by a number of random variations or by improbable and blind molecular or genetic mechanisms.

In fact, as I try to explain in my essay *'Evolutionary Plasticity. A new evolutionary hypothesis based on quantum biology and holographic entanglement'*, the most reliable theory nowadays - that of Darwin - that is based on three fallacious assumptions (the partial ambiguity between the terms *evolution* and *natural selection*, the inaccurate definition of species and, above all, the claim that casuality is the source of mutations), should not be considered valid, and its range of competence should be reduced going from 'theory of evolution of species' to 'ecological theory of races'.

In this context, the *Evolutionary Plasticity*, taking naturalistic observations (mimicry, collective mind of social insects, evolutionary convergence, plants neurobiology, etc.) as a starting point, proposes an 'intelligent and conscious' evolutionism which directly involves the mental faculties of living species, and finally tries to reconcile the evolutionary and creationist positions by means of philosophical considerations which remind Giordano Bruno's pantheistic monism.

This is not an easy task, but *quantum biology* and *holographic entanglement* may provide us with some useful tools to understand, or even just to hypothesize some mechanisms of interaction between mind and body and between the individual and the universe; interactions that some philosophers and mystics of the past as well as some present-day scientists have considered possible.

The theory explicitly refers to the quantum concepts of Bohm, Aspect, and Pribram (holographic paradigms).

11

My ideoplastic theory also introduces new concepts of *hypergenomic complessione* and of *hyperspecies*, which, in my opinion, might explain the appearance of single evolutionary steps and suggest a point of contact between the theory of punctuated equilibrium of Gould and Eldredge and the gradualism.

This naturalist essay does not intend to be exhaustive; the only purpose of this essay is to propose an alternative vision of the evolution issue that takes into account the possible correlations between the alleged 'ideoplastic' phenomena (mimicry, hypnotic ideoplasy, monoideism) and the evolution of living species, as well as possible interactions between the genes and the psyche. Everything in the light of findings and hypotheses that come from the related fields (such as quantum physics, hypnology and plants neurobiology), and in the light of philosophical speculation that is perhaps more enlightening.

All this was done in the hope that my theoretical proposal could stimulate future studies and multidisciplinary researches that help us to understand the process that led to the formation of the leaf-bug and consequently, of the mechanism that caused the appearance of evolutionary mutations on living beings.

(The author)

Kallima inachus (leaf butterfly)

1. The starting question: how has the *leaf-bug* evolved?

'What if Darwin was wrong?' I burst out as I watched some leaf-bugs in a showcase that was on a table in front of me.

I was in the entomology lecture hall of the Faculty of Science and Agricultural Technology housed in the wonderful eighteenth-century building of the Bourbon Royal Palace of Portici, near Naples, to complete my preparation for the agricultural entomology test I wanted to take before the summer holidays. I alternated gaze between the leaf-bug and a colored rainbow of light projected by the glass window on the extremely tight beige short of a colleague elegantly bent forward on a bulletin board a short distance ahead, while the cool breeze of the Gulf crept through the half-open shutters of the old veranda carrying the scent of the ocean and the playful chirp of the seagulls.

'Wrong about what?' asked a very polite male voice that came from behind.

I turned around and swallowed in surprise. It was Professor. Ermenegildo Tremblay, an eminent entomologist and discoverer of the Trioza tremblayi. He had the noble bearing of a lord and,as usual,was wearing the white unbuttoned coat over the suit.

'In thinking that evolution is caused by random mutations...' I replied stammering.

'Ah!' said the professor while scutrinized me from behind his round glasses with his smiling and mysterious eyes. *'You must study, study. And do not let yourself be distracted by Darwin... and by beautiful girls'* he added with a friendly smile turning his eyes on my colleague.

The girl, who had certain resemblance to the Barbie doll, turned around and greeted the professor respectfully. Our eyes met for a moment and she blushed slightly.

I did not follow the professor's advice and I did not take that exam until much later. I understood many years later that, at that time, there were already many elements of what later would become the hypothesis of 'Evolutionary Plasticity' and also of the fantasy story I narrated in the fiction-thriller 'Metamorfer. The gem of Darwin'.

But, let's go back to the leaf-bug.
The body of these amazing insects is strikingly similar in shape and color to the underside of the leaves. It shows a number of

13

marks similar to the ribs of vascular bundles and some color shades that perfectly simulate the stains of dry leaves. Even their eggs mimic the seeds of herbaceous plants.

They belong to the *Phasmatodea* class as well as stick insects.

The term 'Phasma' comes from the greek and means 'ghost'; it indicates that these insects merge so well into the foliage that are no longer distinguishable(crypto-mimicry).

Those who had never seen them live, can get an idea by searching the terms leaf-insect and leaf-bug, or the names of the most representative species (*Phyllium giganteum*, *Phyllium bioculatum*, *Phyllium pulchrifolium*, *Phyllium philippinicus*, *Phyllium jacobsoni*, *Phyllium ericoriai*, etc..) on a search engine.

These are oviparous and often parthenogenetic insects. Furthermore, they are capable of partially regenerating the appendages lost, such as legs and antennas.

So, watching them carefully, I realized that their strong resemblance to the leaves they mimicked could not be completely explained neither by Darwin's theory of evolution nor by any other theory that I know, and therefore, it was necessary to find new explanations for the mechanism that made them evolve like that.

It seemed clear to me that the mechanism that leads to the other kinds of evolutive adaptations should not be very different from the mechanism that triggered those mimetic phenomena; and also convinced myself that it should be a psychic mechanism.

But, let's begin at the beginning.

First, let's analyze the reasons that - in my opinion - make the other evolutionary theories unacceptable, and then deal with the aspects that connect evolutionism to mimicry.

2. The inadequacy of Darwin's theory.

It is common knowledge that Charles Darwin's theory of evolution is based on the concept of *natural selection*, that is to say, on the belief that environmental and ecological mechanisms favor the survival and spread of individuals with the fittest physical, physiological and behavioral characteristics for the environment they live in.

Such mechanisms are real and undeniable, but I would like to make some clarifications.

The first clarification is that the concept of natural selection was not introduced by Darwin, it was introduced by Arthur Russel Wallace, who first presented it in three letters he sent to Darwin and then in the article published in 1858 under the title: *'On the tendency of varieties to depart indefinitely from the original type'*. By mentioning this piece of news, which is now considered a historical certainty, I do not intend to belittle the importance of the studies carried out by Charles Darwin; I just want to emphasize the inexplicable ideological conformism that invaded part of the academic world making it defend Darwin's positions at all cost, and deny the facts for nearly a century and a half.

The second clarification is that natural selection can not be the only explanation of the evolutionary mechanism: in fact, as has already been mentioned, natural selection eliminates the 'less fit' and favors the 'fittest' but, of course, it can only be done when both are present.

According to the neo-Darwinists, how were these mutated individuals born?

Well, the response provided by some scholars (and by the popularizers who have undeservedly enriched themselves by exploiting the controversy around a theory to which they made no contribution) is that evolutionary mutations occur by casuality!

And that is exactly the main flaw of Darwin's theory. Attributing evolutionary mutations, many of which were extremely complex, to the random effect of the blind chance is not a very 'scientific' attitude!

Personally, I consider it 'non explanatory'.

The most arrogant neo-Darwinists even declared, after misinterpreting some experiments, that the fortuitous nature of mutations has been already proved in a number of laboratories around the world!

That is not completely true and we will talk about it in the chapter dedicated to the SAMs, to conformism and genetics. Reintroducing the initial question, I just want to make something clear: on one hand, it is acceptable that natural selection could favor the spread of the leaf-bug, but on the other hand, it is completely unacceptable the hypothesis that they were formed by chance.

When I presented my hypothesis of a possible link between mimicry and evolution (assuming that both mechanisms are caused by the same ideoplastic force), the SAMs, after smiling at me as you normally do with children, reminded me the example of the industrial melanism and of the *Biston betularia* (peppered moth).

And I, after smiling back with the same arrogance, asked them what was the connection between this example and what happened to the leaf-bug, but I received no satisfying reply; as a matter of fact, I realized some of them did not even know these amazing mimetic insects existed.

The *Biston betularia* (peppered moth) is a moth that belongs to the family Geometridae, which larvae feed on leaves of birch, beech, willow and elm. There are three chromatic forms: the clear phenotype (with white wings), the melanic phenotype (with black wings) and an intermediate phenotype (marbled) that manifests itself in case of heterozygosity.

The phenotypic trait 'color of the wings' is controlled by a single gene, in which the *carbonaria allele* (black wings) has almost complete dominance over the *typica allele* (white wings).

Well, it is clear that before the industrial revolution in England, most of the *Biston betularia* had a clear phenotype; then with the pollution caused by industrialization, the black phenotype started gaining ground. This phenotype increased the fitness of the species as it enables them to mimic birch trunks blackened by pollution in industrial areas. So, while black moths escaped birds' predation, white moths were more easily detected and eliminated.

Therefore, this example confirms the effects of natural selection on the distribution of a certain allele in a population. And, in this case, it is a simple trait that can be mutated even by chance (since the main difference between the two phenotypes is that one produces a certain pigment and the other not).

But it must be clear that if we focus the investigation on the appearance of the leaf-bug, we will be dealing with a completely different level of complexity.

In fact, how many genes must have changed the leaf-bug to assume that particular form of the borders of the body that is identical to the lamina of the leaves? How many genes must have changed the leaf-bug to make appear the ribs resembling leaf veins at the right place? How many genes must have changed to obtain stains that make it look like a dry leaf? And how many genes to make its eggs look like seeds of herbaceous plants?

And what are the chances of all these variations occuring simultaneously and making them look like a leaf?

In fact, as it has been highlighted by the famous Haldane's dilemma, the probability that such a complex mutation had been caused by chance is derisive enough to be considered almost null.

However, advocates of random mutations do not give up so easily: they assume that evolution is the result of a number of slight mutations that joint over time, and that natural selection has always discarded the unfit ones.

Does it mean that during the evolution of the leaf-bug, there was a phase in which, for example, the ribs were spiny, or pointed, or hatched? Or does it mean that the insect's body was square, pentagonal or star-shaped? Was it red or blue or yellow instead of green? Does it mean that some right mutations took place after the occurrence of a number of wrong mutations? Does it mean that these right mutations joined the other right mutations instead of the wrong ones?

In my opinion, this second hypothesis seems even more unbelievable than the first one. In fact, this one greatly multiplies the number of 'attempts and errors' that nature should have completed to finally be able to create the leaf-bug.

And while leaf-bugs were undergoing this long series of partial and slightly mimetic mutations, how come their natural enemies did not prey them and eliminate them from the planet? Were they tossing a coin, by any chance?

I would like to point out that the randomness of the appearance of adaptive mutations is unacceptable both from the statistical and from the conceptual point of view.

What I am trying to say is: why would the leaf-bug precisely adopt the form of a leaf of its environment instead of the form of a banana, potato or *Cannabis* leaf?

This is the answer of the neo-Darwinists: because if it had adopted the form of a different leaf, it would have been discovered and preyed.

But this response is not quite valid: in fact, entomophagous insects and birds do not eat banana, potato, or *Cannabis* leaves (unlike some SAMs that I know who, judging by what they say, seem to give them a non advisable use). Therefore, it is evident that the question has not yet been satisfactorily answered.

But I would go even further and ask myself: how come we have never found an iPhone-shaped insect? Why are there no animals with the Google or Facebook logos on the back?

Yet, chance could have also produced animals with those inscriptions. Certainly those animals would not have been attacked by their predators (I know from my own experience that: no insect or bird has ever attacked my iPhone or the screen of Google), and thus, would not have been victims of natural selection and would still exist in nature!

In short, if chance acts freely, should not it create the most bizarre and improbable forms?

And why do we only find leaf-bugs and no Google-bugs in nature?

It is simple: we find the leaf-bug because an ancestor decided to precisely adopt the shape of a leaf present on its natural habitat, or because his mind was stimulated by everything in its environment! Google, Facebook and the iPhone did not exist yet, and therefore, could not be mimicked.

The basic idea of the theory that I called *Evolutionary Plasticity* and that proposes an *ideoplastic Evolutionism* has its origin in this precise certainty. That is: the occurrence of evolutionary mutations is not a random process, it is caused by the will of living beings and responds to precise environmental stimuli.

Nothing hinders that it can be carried out *gradually* (when mutations take place step by step in more than one generation) or *completely* (when the adaptation is completed in a single generation).

The only difference is that in the first case, due to the persistence of the state of necessity and to the partial adaptation obtained, the ideoplastic action would be exercised by the minds of partially mutated individuals of several generations.

3. Considerations on Lamarckism.

Lamarckism, which precedes Darwinism, has the great merit of introducing the concept of evolution as opposed to that of fissism which stated that living species were immutable.

Furthermore, although it has been largely refuted over the years, I believe this theory still preserves a conceptual structure that is worthy of the highest consideration. I am referring to the assumption that evolution is not the result of chance, but of an evolutionary necessity - and I fully agree with that.

Two aspects of this revolutionary theory were questioned: the presumed mechanism that induces mutations and the fact that fails to explain how the somatic characteristics acquired are transmitted to future generations.

Regarding the first aspect, it is known that, according to Lamarckism, the appearance of new characteristics on cells and somatic structures is induced by a physical effort made by an individual (through mechanical stimuli and transference of "fluids" to the most stimulated organs, while the disappearance of certain appendages was due to the fact that they were no longer used or stimulated (principle of *use and nonuse*).

The typical examples were those of the elongation of the neck of the giraffe (that developed a long neck because of the physical effort made to reach the leaves on the top of the tree) and of the snake (that lost its legs because did not use them).

But some scholars, including Georges Cuvier, the founder of the vertebrate paleontology, opposed this theory because it did not explain the appearance of features that are not conditioned by use and nonuse of members, such as the punctate mimetic skin of the giraffe (and I would also add all the mutations with mimetic purposes mentioned in this informative essay).

Other opponents recoursed to a sadistic experiment to question both, the validity of the principle of *use and nonuse* and the theory that the somatic characteristics acquired could be transmitted to subsequent generations. They cut off the tails of some laboratory mice (which, thus, could no longer use them) for several generations to prove that the descendants were always born with tails: therefore, variations in the parents' body caused by physical factors were not transmitted to their offsprings.

However, I share Lamarck's idea that the appearance of new

features in a living being is not caused by chance, it is caused by the need of the individual to respond to environmental pressures or to adapt to new living conditions.

The substantial difference between Lamarckism and *Evolutionary Plasticity* (or *ideoplastic evolutionism*) lies in the fact that, while according to Lamarck the individual induced the appearance of new features through physical actions (*use and nonuse*), and in my opinion the individual is able to induce adaptive mutations acting - through psychic and ideoplastic actions- directly on the genome of his own germinal cells.

Therefore, my hypothesis could explain the appearance - in a short period of time - of new desirable characteristics (and that do not force chance to make billions of attempts and errors, as neo-Darwinists want), and the appearance of features that are not related to the principle of *use and nonuse*, and could also explain the method of transmission of new features to the descendants.

So, according to my hypothesis, the elongation of the giraffe's neck was not caused by the physical effort made by the animal (what could have only caused a minimal effect on the individual that tried to reach the leaves on the top, but certainly would have had no effect in the descendants), it was caused by the desire and the *ideoplastic action* of the individual's mind that induced the necessary variations directly in the genome of its germinal cells and, consequently, in individuals of following generations using the methods I will present later.

And that applies not only to the elongation of the giraffe's neck, but also to the mobility of the penis of the elephant (male elephants have developed a penis that is mobile just like the proboscis and that touches the abdomen of the female searching the right way) and to the evolutionary mutations of every nature and of every living species.

I should add that all the mutations described in the following chapter and that regard evolutionary variations with mimetic purposes, can not be explained by means of the Lamarckian principle of *use and nonuse*, but can be easily explained by the *Evolutionary Plasticity*.

4. Relationship between mimicry, evolution and evolutionary convergence.

When I was studying zoology, I observed a particular behavior of the *sepiidae* that made me think.

These animals are very intelligent marine mollusks, commonly known as the cuttlefish, that belong to the *cephalopod* class. The feature that impressed me the most is their strong mimetic abilities which they used in extremely creative manners. For instance, sometimes some young males adopt the appearance of sexually immature females to escape the surveillance of bigger males and be able to copulate with their females.

But, even more important, these cuttlefishes manage to instantaneously adjust the color of their bodies to that of the substrate on which they lay (camouflage).

This phenomenon is called *rapid-mimicry* since it is an instant and temporary reaction some how similar to the blush on human beings.

Well, I hypothesized that this is not an involuntary phenomenon, on the contrary, it is a voluntary reaction caused by an *ideoplastic somatization* (the action of the psyche on the body) of the same type (crypto-mimetic) experienced by leaf-bugs, but this time managed to permanently fix the similarity to the leaves in its genes (*genomic imprinting*).

I believe that mimetic phenomena, whether *rapid-mimetic* or *crypto-mimetic*, constitute a clear proof of the effective influence of the psyche on the body and on genes, and that the mechanism that induced them is very similar to the one that causes the evolution of all living species.

According to my hypothesis, the mind would be able not only to affect the individual's physiology and control the expression of genes, but also to create new genes.

Everything that has been hypothisized till now has not yet been proved, but can be a fertile field of investigation.

I, for my part, will try to overcome the main objections to my evolutionary hypothesis in the following chapters by presenting some clues that may support it and suggesting some possible explanations for the mechanisms involved. I invite all those who want to study these aspects in detail to read my other publications about this topic.

For now, I would invite readers to observe mimetic individuals without prejudices because the careful study of this animals may help us to understand the entire mechanism of evolution.

For example, if we observe the *Kallima inachus*, of which I recommend you to search some images on the Internet, it is inevitable for us to be surprised by its appearance and to ask ourselves some challenging questions.

The upper part of the wings of this butterfly has gaudy colors that are useful for courtship, but if it foldes its wings flattening them vertically one on top of the other, you can see the coloring of the lower side that is similar in every possible way, even in shape, to that of dry leaves typical of the habitat in which it lives.

Some other animals that should be carefully observed are the *Uroplatus phantasticus* or leaf-gecko; the 'common sea dragon' (*Phyllopteryx taeniolatus*), the seahorse that modified its body to resemble the algae of its ecosystem; and the *Aegeria* or *Sesia apiformis*, a butterfly that resembles a bee to escape predators (phoberic mimicry).

Well, it seems clear to me that the evolution of these mimetic species, as weel as that of the many other species that we find in nature, was admirably 'oriented' and that can not be produced neither by a *random mechanism* nor by *use or nonuse*.

And this also applies to the evolution of leaf-bugs, giraffes, and of all the nonmimetic animals.

For example, it is known that all the animals that live in a certain type of environment end up having the same shape and the same functions.

The ichthyosaurus (primitive aquatic saurians), the cetaceans (dolphins, orcas, whales, porpoises, etc..) and fishes have a very similar hydrodynamic shape and have developed fins.

Since both, the bat, which is an insectivorous mammal, and the guacharo, a strange bird with oily feathers, live in dark caves, developed a kind of sonar to orient themselves.

The common mole and the marsupial mole, apart from the difference in the color of the hair (the first one is black and the second one is yellowish), both have the same shape even though the first one is a mammal and the second one is a marsupial like kangaroos*.

These species are called 'convergent' and the evolutionary boost that causes the appearance of common features is called 'evolutionary convergence'.

---

*The mole-cricket (*Gryllotalpa gryllotalpa*), which is an insect that lives in the soil, has its body similar to that of the moles.

Well, according to the theory of the *Evolutionary Plasticity*, since all these species undergo the same type of environmental pressures, they end up acquiring the same adaptations. Therefore, they undergo the same transformations (*ideoplastic evolution*) and end up strongly resembling each other.

Of course, it is possible that natural selection subsequently acts on the heterogeneous population present and favors the spread of the fittest individuals. However, it is clear that it can not be considered the primary cause of evolution, it is just one of the factors that completes it.

The ideoplastic mechanisms by which the mind of living beings induces the appearance of evolutionary mutations and produces new genes are not quite clear, and we will try to hypothesize the possible mechanisms in the following chapters.

But the fact that we have not understood yet how the ideoplastic mechanism of mimetic mutations works does not mean that it does not exist. The leaf-bug exists, it is identical to a leaf and there must be an explanation for its appearance! Once we figure out the nature of the mind-genes interface, we will be able to understand the general mechanism of the evolution of all living species!

But first we must deal with an issue of paramount importance: evaluating the credibility of the hypothesis that all living species, including plants, possess a mind.

5. Does every living being have a mind?

In short, my opinion coincides with that of the Nobel Prize winner Max Planck, who is considered the father of the quantum theory, which states that: «*All matter has an origin and only exists by virtue of a force... we must assume that there is a conscious and intelligent mind behind this force. This mind is the matrix of all matter*».

And I would not be at all surprised if it turns out that all systems with a flow of energy (including stars and plasma) manifest phenomena of consciousness and intelligence. Furthermore, considering that matter is just one of the many forms of energy, it would not be hard for me to believe that it is intelligent too.

Obviously, at this moment we have no proves of that, but some open-minded researchers are also investigating in these related fields.

Some of them, for example, are trying to determine if computers have consciousness (*eg, Can Machines Be Conscious?, Koch and Tononi G., in "IEE Spectrum", Vol 45, No. 6, pp. 54-59, June 2008*).

In subsequent chapters, I will refer to some philosophical and quantum aspects related to such controversial subject, because in this phase I would like to focus on naturalistic considerations only.

In particular, I want to emphasize that, in my opinion, it was the belief - rather naive than erroneous - that animals, plants and microbes do not possess any mental faculty what prevented scholars of the past from daring to hypothesize that the evolutionary process could be triggered by a psychic factor, as I openly suggest.

Consequently, these scholars have had to find alternative mechanisms of induction: Darwinists tried to present *chance* as the cause of the evolutionary boost, lamarckists tried to explain it with the physical principle of *use and nonuse*, supporters of Intelligent Design tried to explain it through the direct action of God, and new-age scholars through improbable automatic interactions of different natures.

Yet, Darwin himself observed in '*The power of movement in plants*' that radical apexes of plants functioned as an extended brain, similar to that of lower animals. But, obviously, not even him risked to make a fool of himself by saying that plants could be conscious and have a real mind. Indeed, in more recent days, some observations aimed at proving the alleged psychic abilities of plants (the Kirlian effect, observed in 1939, and the experiments carried out by Cleve Backster in 1966), were heavily opposed and derided by mainstream science.

For my part, I found it easy to believe that animals and microbes possess remarkable mental abilities. And in my opinion, it has already been proved by the simple fact that they interact with the environment, feed, reproduce and have survived for millions of years, and do it with undisputed success despite the observations - very hilarious at times - of some scientists and of the usual SAMs.

I notice how surprised and amazed are some researchers at finding, for example, that a monkey manages to take a peanut that is inside a glass by raising the water level in it. I, on the contrary, am not at all surprised at the intelligence shown by the monkey (which to survive has to solve much more complex and vital problems in nature), but I am seriously doubting the intelligence of some researchers and am sure they are wasting their time and taxpayers' money.

As for the alleged intelligence of plants, I explicitly refered not only to Darwin's observations mentioned above, but also to the studies carried out by prof. Stefano Mancuso from the University of Firenze and to his research field called 'plant neurobiology'.

In the end, I admit that no final or acceptable conclusions have been reached on these aspects. However, I want to highlight that mimicry phenomena occur even in plants and that can not be considered random. It makes me think that may be plants used their mental abilities to adopt certain shapes.

I am referring, for example, to the particular shape of the *Ophrys apifera*, an orchid that transformed part of the flower making it identical (in color, shape, size and odor) to the abdomen of a female bee in order to attract male bees and help pollination.

So I would urge researchers not to ignore the naturalistic observations presented by me, to seek evolutionary explanations other than casuality, and to reintroduce old hypotheses - like those presented by Kirlian and Backster (theory of *primary perception*), perhaps too hastily discarded - and eventually riconsider them in the light of the new perspectives provided by quantum physics and by the observations related to the presumed bio-entanglement, some of them presented in my previous essay.

Neurosciences are advancing by leaps and bounds and we can not ruled out the possibility that researches on BCI (*brain-computer interface*, also known as MMI, BMI, or direct neural interface) can be partially used by more inspired researchers to study the *ideoplastic effect* of the psyche on the production of new genes, as I hypothesized.

6. Relationship between mind, genome, hypnology and quantum sciences.

As it has already been said, the hypothesis that the mind can affect the body and the genome causing the appearance of new genes, is the main basis of the theory of *Evolutionary Plasticity*; and it reaffirms that the very same fact that rapid-mimetic organisms are able to change the appearance of their own bodies is a clear indication of the ideoplastic power of the psyche.

I provided some evidence and considerations aimed at strengthening my hypothesis of study in the essay *"Evolutionary Plasticity. A new evolutionary hypothesis based on quantum biology and on the holographic entanglement"*, and refer you to it for further reading.

Here I will only give a few examples from the fields of medicine, psychiatry, and hypnology.

I remember, above all, that both the *placebo effect* and its opposite, *the nocebo effect* are usually considered real in medicine. They clearly prove that beliefs and the mental state of individuals can significantly condition their health.

Even psychiatry provides us with an interesting example of the influence of the mind over the body: the MPD or *Multiple Personality Disorder*.

When people suffering from this condition believe to have a certain personality, they present some physiological symptoms (for example, are violently allergic to insect bites) and, when they believe to be another person, their behaviour is completely different (for example, show no or very limited allergic reactions). It has also been observed that, in some cases, there is even a variation in the color of the iris of their eyes.

These behavioral differences have no genetic basis since we are talking about the same individual with the same genetic makeup. These differences have a psychic basis that can only be explained by admitting that the psyche of an individual can deeply condition his physiology, and may be even interact with the mechanisms of epigenetic control.

Other impressive effects of somatization, sometimes associated with MPD, consist of the hysterical pregnancies and perhaps - at least in some cases - by the stigmata and self-healing miracles. In particular, as regards the latter, it is not to exclude any type of interaction with the stem cells.

Hypnology also provides some examples of the possible control exerted by the mind, especially of its unconscious component, on the body.

In fact, there are a number of cases presented in literature that prove how a subject in a state of hypnotic trance is able to control the pain, to get rid of the warts and even increase the circumference of the breast.

In this respect, I present an abstract of a study carried out by Dr. Willard (*Willard, RD, Breast Growth through visual imagination and hypnosis. "The American Journal of Clinical Hypnosis", 1977, 4:195-200*): «After 12 weeks, 28% had reached the objective fixed at the beginning of the program and wanted no more growth. 85% reported a significant growth of the breasts, and 46% found it necessary to buy larger underwear. The fourty two percent lost over 4 pounds of weight and also experienced abreast growth. The average increase of the circumference was of 1.37 inches; there was a vertical increase of the measures of 0.67 inches and a horizontal increase of 1.01 inches».

However, all these examples only prove that the mind can affect the physiology and some somatic features (somatization), but it does not prove the existence of a direct relationship between mental action and the modification of the genetic inheritance. Yet, if we accept that the mind can influence somatic cells, it is perfectly possible that it also influences germinal cells and DNA.

I am certain that the proof of the interaction between mind and DNA is already in front of us and that is easily recognizable if we look at it without preconceptions.

And I am referring to some episodes of 'somatic imprinting' that I presented in my previous essay (such as the kittens born with a date or with the word *cat* on their bodies, and a hen that laid eggs in the shape of nuts), to the particular shape adopted by leaf-bugs and by their eggs, to the giraffe's mimetic coat and to the alga-shaped body of the weedy sea dragon.

As I said before, according to my *ideoplastic hypothesis*, the mind, in response to a need of adaptation or to an *eidetic visualization* (insight), fixes the mutations in the genes of gametes which transmit them to subsequent generations.

But this is just a hypothesis that still has to be proved. That is why I urge researchers to seriously consider the fact that the evolutionary induction may not be a random event, on the contrary, *it may be an event caused by the mental power of individuals.* I also urge them to perform experiments in order to confirm or refute this hypothesis, bearing in mind that it seems to be supported by my naturalistic observations on mimetic species.

I would particularly suggest an approach that takes into account the results of quantum sciences. I do so because it is a matter of 'intellectual coherence', that is to say, the conclusions reached by quantum physics about the inanimate matter must also be a valid basis for the biological matter, which has the same basic structure.

So, if the matter has a dual nature (corpuscular and ondulatory), the biological systems must be studied also taking into account quantum aspects.

And, if some conclusions are considered valid in the inanimate field (eg, the influence of the observer through the 'collapse of the wave function', the tunneling, the Einstein-Podolsky-Rosen effect, the entanglement, the non-locality, etc..), must be equally valid when studying biological phenomena.

In conclusion, without ruling out other possible interpretations or interactions (possible relations with proteomics, alleged mnemonic properties of water, etc..) and taking into account the considerations that I briefly present below, I am inclined to believe the hypothesis that the mind-body and mind-genes interfaces can be ondulatory, and I make direct reference to Bohm and Pribram's holographic paradigms.

Karl H. Pribram, distinguished Austrian neurosurgeon, stimulated by Bohm's quantum theories, formulated the theory of an holographic model of the brain (Holonomic Brain Theory) in which information and memories are not registered in neurons; they are the result of wave patterns (or interfering pattern) represented by Fourier's equations, and in this way explains the brain's ability to store a huge amount of information in a relatively reduced space.

So I, supporting Bernstein's hypothesis, believe that not just the brain but the entire body is holographically connected. Therefore, all the information, including the memory, would be distributed and stored throughout the body in anondulatory field of quantum nature.

I support this hypothesis based on to three considerations: the alleged memory of the organs of transplant patients, the possible implications of Valerie Hunt's experiment, and the alleged collective mind of social insects.

In fact, many people who have undergone a transplant of one or more organs, acquired some of the donors' habits without having previously received information about these habits. Everything indicates that the transplanted organs preserved the donor's memory through a mechanism still unknown, which I believe to be holographic or undulatory.

Valerie Hunt's experiment makes you think there is an energy and immaterial field that surrounds the human being.

This American researcher was studying the responses of some subjects to a light stimulus and comparing the readings of the electromyograms (EMG) and electroencephalograms (EEG) when observed that, unexpectedly, the electromyogram registered a faster response to the stimulus than the encephalogram. This fact would prove the existence of a mental energy field that surrounds the body and even supervises the brain functions.

And the idea that the mind is formed by an energy field separated from the biological substratum but at the same time connected to it (bio-entanglement) can also be reinforced by observing the behavior of the so-called 'social insects' (ants, bees, termites).

The colonies of these insects are often called superorganisms due to the fact that their behavior seems to be coordinated by a 'collective mind' which leads them to do things so complex that can not be explained only using chemical (pheromones) or sign language.

On the other hand, the idea of the existence of morphogenetic fields with the capacity to shape the form and functions of a developing individual is not new and was suggested by biologists in the 20s of last century, and has been more recently echoed by other scholars among which Rupert Sheldrake certainly is the most prominent one.

Finally, I want to emphasize that the idea of the existence of a *field of force* around the human body, more or less intelligent and usually called *aura*, is very old: in India it has been called *prana* for over 5,000 years, in China it is called *ch'i* and the Jewish Kabbalah calles it *nefish* and describes it as anoval-shape iridescent bubble that surrounds the human body and is visible to some mystics.

I repeat, despite *Evolutionary Plasticity* is mostly based on naturalistic considerations (specially on the relationship between mimicry and evolution of species), the direct reference to Bohm's holographic paradigm may be useful to answer a specific question: when the results of a given evolutionary mutation is not yet present in nature, where does the individual's psyche obtain the necessary information from? The answer is: from the area of the physical universe in which the matrixes of these forms reside.

I know: this answer will make rejoice all the SAMs of the planet who consider it fanciful enough to almost border on idiocy.

However, this idea did not seem ridiculous to many philosophers (Giordano Bruno and Hegel's philosophy, Plato's Hyperuranium,) and cognitive science scholars (eg, Jung's collective unconscious). It did not seem ridiculous to many theoretical physicists that, not only in cosmology but also in quantum physics, presented similar theories suggesting that the objective reality, with four dimensions (three spatial dimensions plus time), is the holographic projection of a bidimensional matrix located elsewhere.

Bohm has particularly theorized the existence of two physical non-local planes holographically connected: an *implicate order* in which resides the holographic information of everything that exists in physical reality, and the objective reality which he called *esplicate order*.

And Albert Einstein himself expressed an analogous concept with the phrase: "*Everything is determined by forces over which we have no control. It applies to insects as well as to stars. Human beings, vegetables or cosmic dust, we all dance to the rhythm of a mysterious music played in the distance by an invisible piper*".

I must add that Bohm's implicate order, which is very similar to Plato's *world of ideas*, could be the theory in which reside not only the physical matrixes of living beings but also the matrixes of their instinctive behavior, which could explain the astonishing behaviors of some living beings (for example, the incredible skill with which spiders spin their webs and bees construct the hexagonal cells of their hives).

At this point, I admit that many of the observations presented in this chapter (such as the ideoplastic effects and the somatization observed in hypnology and in psychiatry) do not take place so frequently and sometimes are hardly repeatable, but it does not mean that they do not exist and that should not be studied.

Moreover, I emphasize that evolutionary events are not so frequent either. For example, human beings have not evolved at all in the last milleniums.

Therefore, we are dealing with infrequent phenomena and, in my opinion, that is another important feature that psychically induced phenomena and evolution have in common.

This may happen due to the fact that some particular and extreme conditions are required for them to take place. These conditions are equivalent to *monoideism* in hypnology and to specific and intense states of necessity in the evolutionary field, that in some cases even need to be prolonged for some time.

At this point, I want to insist on two basic concepts.

The first one is that searching for possible explanations for the evolutionary phenomenon even among the most imaginative hypothesis would be always better than settling for a 'non explanation' represented by the sterile, dogmatic and unscientific hypothesis of the alleged random mutations.

The second one is that if some conclusions are accepted in one research area, must be also accepted in the other research areas (what I call principle of 'intellectual coherence'): so, for example, I see no problem on applying some principles of quantum physics to the study of the evolution of living species.

Finally, I would like to point out another possible element for reflection: all the phenomena in nature are regulated by precise laws, forces and rules. Why should the evolution be the exception? Just to please the SAMs and the neo-darwinists?

7. SAMs, *Podaciris sicula* and relationships with genetics.

The SAMs (Middle Academic Scholars) are an extremely peculiar anthropological group which main characteristics are egocentrism, presumptuousness, rudeness and irascibility.

They are often anonymous researchers who have never discovered anything relevant and therefore, are quite frustrated for having spent so many years of their lives doing nothing.

Sometimes they even hold a university chair. Other times they write books on subjects on which they have made no intellectual contribution and, although undeservedly, earn lots of money.

In this last case, they defend their small field of interest with particular vehemence like those dogs that protect their limited and vital territory to the death.

Even when they generally have a limited and sectorial training, and often different from the subject they are writing about, they believe to be experts in every branch of knowledge and deeply offend (either in the titles or in the content of their books) anybody whose opinion differs from theirs.

I will mention no names for now, but they are easily recognizable by their fundamentalist and scientistic positions.

They are, metaphorically, like donkeys with blinkers (accessories that force them to see just part of the world around them) and often even have problems with the cervical pathology (that does not allow them to turn the head, not even a little, to change the field of vision), and are myopes too (because they can not see further than the ends of their noses). They are extremely conformist (see, in this respect, the enlightening experiment on conformity carried out by Solomon Asch in 1956) and tend to deny every evidence opposed to their rigid convictions or to their own interests, and defend the dominant ideologies to death.

And they are hardly as nice as creationists (which, at least, are not presumptuous enough to believe only in themselves)!

In short, even though unintentionally, they seriously slow down the scientific progress by hampering the study of everything that has not been discovered yet, and do not realize that such attitude will lead mankind to make no further discoveries!

In addition, they treat all those people whose opinions are different from theirs with an air of superiority, with presumptuousness and rudeness.

Creationists, on the contrary, are much more sympathetic and, even though I do not share their views, must say that for *par condicio*, their opinions also deserve a modicum of consideration.

As pure speculation, even when there are many intermediate fossils that support the evolutionary theory and geology has proved that the Earth is about 4.54 billion years old, they could maintain that the Earth is only three days old and that fossils do not prove anything.

In fact, if the Earth has been actually created by a God, it could have happened just three days ago and it may have been created already having '4.54 billion years of age'. And, again three days ago, this God may have scattered some fake fossils here and there and implanted in our mortal minds some memories (taxes and loans included) of what we believe has been our whole life till now.

All right: I admit that this theory is pretty stupid.

But, if we want to be objective, it is not much more stupid than that of supporters of random evolution who, being familiar with lies, state there is evidence that the evolution of living species is caused by the chance in many laboratories around the world.

I would like to say to the SAMs that if they have ever seen a hairy and primitive-looking female coming into a lab and then a beautiful and fascinating modern woman walked out, they should check it: because that laboratory must have been a beauty center instead of a genetic one!

Irony aside, I know - as well as the SAMs do - that through the use of genetic engineering we can make living species acquire new and interesting features. And I point out that, as an agronomist, I have studied many researches about it, the first ones were carried out on the very same field of my expertise: that of vegetable organisms and animals of zootecnical interest.

However, it must be clear that all the variations obtained in laboratories are not random at all (since have been intentionally produced by the man) and that slight experimental or artificial variations have never led to any significant and functional mutations comparable to the appearance of a new species or of a new functional organ.

The random genetic mutations may lead to small variations (micro-mutations), often harmful and rarely functional, but that may even lead to the appearance of complex evolutionary mutations, that involve many organs and that must act in perfect harmony, is a theory that still needs to be proved.

33

But, of course, I am always willing to revise my opinion if someone proves me otherwise. In the meantime, in absence of irrefutable proofs, and strongly believing in the evolution of living beings, I maintain that it can not be determined by chance and that is caused by some force that we have not yet understood.

Therefore, based on my naturalistic observations on mimicry, I will continue supporting the validity of the *ideoplastic evolution* or *Evolutionary Plasticity*.

I would also like to make clear that even the example of the *Podaciris sicula*, that some SAMs love, may have other explanations different from the one proposed by them that was based on chance.

It is common knowledge that in 1971 Prof. Eviatar Nevo introduced a number of specimens of a small lizard, the *Podaciris sicula*, in Hrid Pod Mrèaru's Dalmatian islet to study their adaptation to the new habitat.

In 2004, a team of scientists led by Duncan Irschick and Anthony Herrel returned to the island and noticed that the lizard had experienced some morphologic changes.

The mitochondrial DNA tests undergone by the lizards on the islet confirmed that they were members of the species *Podaciris sicula*, and yet presented significant differences: they were larger than the continental ones, their jaws became stronger, their diet had been modified (going from insectivorous to herbivorous), and a new structure had appeared in their digestive system: the ileocecal valve, a favorable adjustment to the new diet.

According to the neo-Darwinists, this would be a clear example of evolution. And I tend to agree with them even though, unfortunately, we can not ignore the fact that male lizards which carry the feature 'ileocecal valve' may have already been in that islet, or may have arrived between the years 1971 and 2004, without us and researchers knowing it. In this case, the latter could have interbred with the female lizards introduced in 1971 and that hypothetical event would not have been detected by mitochondrial DNA tests, since mitochondrial DNA is inherited from the female parent (as a rule, the few male mitochondria are eliminated from the zygote).

However, if this hypothesis is excluded, which is not unlikely, it is not correct to automatically infer that the presumed mutation was caused by random mutations.

In fact, it is probably quite the opposite.

As a matter of fact, it is difficult to believe that a number of random mutations had appeared in such a short period of time (not exceeding 33 years, but perhaps much shorter, since we do not know exactly when the character 'ileocecal valve' appeared), among which natural selection would have chosen the fittest to the new habitat. Another element that makes the occurrence of mutations difficult to believe is that there were no intermediate species in the islet (e.g, with incomplete or partially functioning ileocecal valve) or other wrong mutations.

Therefore, if there was any mutation, it is more likely that had been caused by the psyche of the lizards, stimulated by the forced change in diet induced by the environmental characteristics of the new habitat.

This ideoplastic force could have also acted directly on the DNA of germinal cells, in the extremely limited space of a single generation and with no need to resort to trials and random errors, in three possible ways: either, as hypothesized in my previous essay, by activating a 'genomic complessione' already present in the hyperspecies *Podarcis sicula*, or causing an ideoplastic mutation in the DNA, or accessing the information in the matrix of the *implicate order* presented by Bohm.

Natural selection would have then acted on the heterogeneous population obtained and could have eliminated the lizards inserted in 1971.

Thus, the hypothesis of study that I called *Evolutionary Plasticity* does not deny the close relationship between living species and genetics (and also with epigenetics and proteomics), but considers DNA as the vehicle and not as the source of evolution, as it hypothesizes that the presumed ideoplastic or mutagenic action of the psyche would precisely have the epigenomic complex and the DNA molecule as last 'target', and then would explain its functions according to the methods studied and validated in every genetic laboratory of the world, and that my theory completely accepts.

But let's go back to naturalistic aspects.

I would like to mention the example of the chameleon in order to clarify this subject.

This reptile, apart from being mimetic, has developed an everted tongue almost twice as long as its body to capture its preys. Watching this cute animal hunting, one quickly learns that this character can not appeared by chance: the reptile 'wanted' to develop this organ, trained itself and longed to strike its prey in the distance, and eventually was able to transmit this character to its descendants by modifying the genes of germinal cells, perhaps using a mechanism that can be explained by quantum physics.

The same process also occurred, presumably, in the evolution of the anglerfish (*Lophius piscatorius*). This rather plain but tasty fish modified the first ray of its spine and transformed it into a structure called *illicium*, similar to a fishing rod with an worm-shaped fleshy excrescence which uses to attract fishes in front of the mouth so that can devour them.

This evolution can not be attributed to *casuality* or to *use and nonuse* either. It was probably induced by the desire of the parents. And this, in my opinion, is another proof that the evolution of living species is nothing but a *will that takes shape*.

I want to highlight that there are many other examples of mimetic evolution in nature that should also be studied. I recommend readers to search in the internet images of the orchid mantis (*Hymenopus coronatus*) and of the stick-insect (*Bacillus rossius*, etc..). The latter is similar to a branch and merges into the branches by aligning itself with them. If it is windy, this insect even swings like a branch moved by the breeze. Therefore, it is completely aware of its resemblance to real branches and does everything to maximize its mimetism! And yet, goodness only knows why some scholars are convinced that they can not think!

Finally, going back to the example of the *Podaciris sicula* with the ileocecal valve, I must emphasize that it can, by no means, be considereda new species, because, as far as I know, mutated individuals can interbreed normally with lizards without ileocecal valve and give birth to a fertile progeny.

8. The ideoplastic evolutionism.

In sum, the hypothesis of study called *Evolutionary Plasticity* or *Ideoplastic Evolutionism* suggests that the evolution of living beings is caused by a presumed mutagenic activity of the psyche of living beings that acts directly on the genome and on the epigenome of germinal cells (or, more generally, of reproductive cells).

It is based on precise naturalistic observations and, especially, on the study of mimetic organisms.

It particularly assumes that the mechanism that causes temporary variations in rapid-mimetic organisms (eg, *sepiidae*) is of the same type as the one that induced the variations genetically fixed in crypto-mimetic organisms (eg, *phasmidae*), and also assumes that both have points in common with the mechanism that allows all species in evolution to acquire new features.

It differs from neo-Darwinism since does not accept casuality as the cause of evolutionary mutations and, on the contrary, holds that evolution is caused by the desire of living beings. However, even when redefining it, accepts the validity of the mechanism of natural selection.

It is different from Lamarckism because it holds that mutations are induced by a vital necessity. However, it states that mutations are caused by the action of psychic stimuli (*ideoplastic action*) rather than by physical stimuli (principle of *use and nonuse*). Furthermore, unlike Lamarckism, it is also able to explain the appearance of characters that do not depend on *use and nonuse* and the inheritance of features acquired (by direct action of the psyche on the genome).

It recognizes the fundamental role of genetics in the expression of phenotypic traits, but considers that genes are *the vehicle and not the source of evolution* (it does not accept that random mutations in genes can lead to complex and functional mutations; it states that they only lead to micro- mutations which are often deleterious).

It presupposes that all living beings are endowed with mental functions and, regarding plants, it also refers to *plant neurobiology*.

It suggests that some evolutionary aspects can be explained using *Bohm's holographic paradigm*. And finally, while admitting not knowing how the psyche would act on genes, suggests that the mechanism involved may be of *quantum type* (since the findings of quantum physics related to in animate matter must also be considered valid for the biological systems).

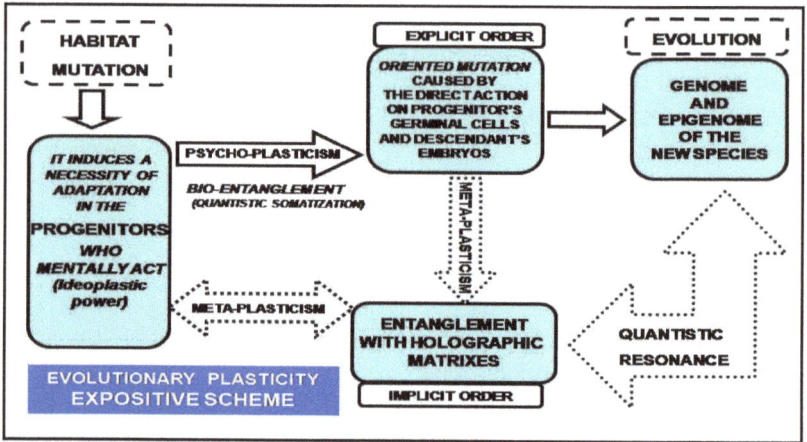

EVOLUTIONARY PLASTICITY EXPOSITIVE SCHEME

HABITAT MUTATION

IT INDUCES A NECESSITY OF ADAPTATION IN THE PROGENITORS WHO MENTALLY ACT (ideoplastic power)

PSYCHO-PLASTICISM
BIO-ENTANGLEMENT (QUANTISTIC SOMATIZATION)

META-PLASTICISM

EXPLICIT ORDER
ORIENTED MUTATION CAUSED BY THE DIRECT ACTION ON PROGENITOR'S GERMINAL CELLS AND DESCENDANT'S EMBRYOS

META-PLASTICISM

ENTANGLEMENT WITH HOLOGRAPHIC MATRIXES
IMPLICIT ORDER

EVOLUTION
GENOME AND EPIGENOME OF THE NEW SPECIES

QUANTISTIC RESONANCE

Once said this, I want to urge researchers to take into account my naturalistic observations and to try to verify (or even to refute) my hypothesis of study, whether with naturalistic observations or with laboratory experiments that examine the possible effects of the psyche on genes. They could also investigate, as I suggested in my novel, the possible effects of the *somatosensory potentials evoked* and of the *hypnotic ideoplasyon* the alleged quantum field that surrounds the body and holographically conditions the DNA.

In fact, I would gladly accept a confutation, because, as Karl Popper said: "Every confutation should be considered a big success not only of the scientist who confuted the theory, but also of the scientist who created the confuted theory because, although indirectly, he also suggested the experiment that refuted it".

I want to make clear that I consider the theory of *Evolutionary Plasticity* as an alternative to the *classical evolutionism* and the *Intelligent Design*: if we admit that evolution is induced by the psyche of living species and that these living species are nothing but particles of God, then stating that evolution is determined by the ideoplastic action of the psyche does not necessarily refute the action of a God that can express himself through the work and demands of his creatures.

From the philosophical point of view, my ideoplastic hypothesis is linked to the *pantheistic monism* of the Dominican monk *Giordano Bruno* (born in Nola, near Naples, in 1548, and burned alive by the Roman Inquisition in 1600 because his ideas were considered heretical), who in his work '*Spaccio de la bestia trionfante*' in 1584, wrote: "Which nature (as you must know) is no other but God in everything."

Archerfish (*Toxotes chatareus*)

According to the *Evolutionary Plasticity Theory* both of these behaviors, as well as the evolution, are voluntary.

Cuttlefish camouflated

The object of contention

(Story by Pellegrino De Rosa)

Giordano Bruno

The object was suspended in mid air, motionless and in silence, over the calm sea.

It had the shape of an isosceles triangle with its base to the left and its tip to the right, and seemed to be metallic.

The child opened his eyes wide, ran to a man called 'Master' Sam, and breathless, tugged at his cassock.

"What is it?"

"Look!" said the child, pointing to the object in the sky. "What is it?"

The man looked up and stood open-mouthed.

The object was huge and scary, like the doubts that gnaw mortals' lives.

"I do not know what it is. How long has it been there?" asked the old man.

The boy shrugged his shoulders and scampered off, troubled: he expected to get some answers from the Master, and got questions instead.

The old man bit his lip.

As far as he knew, that object could have been there, suspended in mid air, since the days of Creation without being noticed. In fact, men were too busy with their own businesses and never raised their eyes to heaven.

The news of the sighting spread in a twinkling and everyone on the island started to walk with the nose upward waiting for the object to move, to go away, to fall or to emit a sound.

But nothing happened!

The object remained in the same place for days, months and years.

Meanwhile, men divided themselves into factions: some were afraid, afraid of the end of the world and formed a religious sect; others began to study a plan to bring it down; others, instead, began to observe it with large telescopes; and others began to adore it as an Idol and succumbed to fornication.

Then, the *Master* had an idea.

He thought that, apart from their island, the island of naturalists, the object should be also seen from the neighboring islands. So, overcoming his innate reluctance, he picked up the phone and called his colleagues who lived in the island of the physicists.

Their answer left him stunned: "We have been watching the object for quite a while, but you are wrong: it is not a triangle... it is a circle!"

"A circle?"

"That's right... and has a point right in the center!" said the physicist.

"Ah, but then you must be talking of another object! Yet, it is strange that we are not able to see it."

"Do not worry, because we see only one object in the sky and again - it is a circle. We do not see your triangle".

"What if you are wrong?" said the naturalist professor. "Perhaps your eyesight is not as good as you think. I could swear it is a triangle!"

"Then let's do something" suggested the physicist, "let's call the religious and ask them what they see."

"Excellent idea. You will see that they will agree with me", said the naturalist.

Then, they called the religious and asked them if they saw something in the sky, suspended over the sea in front of them.

The religious raised his eyes to heaven and fell silent.

It was true!

There was a huge object suspended in mid air in their sky and he had not noticed it till now.

He dropped the phone out of amazement and rang the bells to summon all the inhabitants of his island.

"Had any of you noticed this object in the sky?" he asked.

Everybody shook the head.

No one had ever noticed!

So, he called the naturalists and the physicists and told them: "There is a mysterious object in our sky too, but in our opinion, it has the shape of a squashed circle - although it looks more like an ellipse - with a small triangle on the right. Therefore, it should not be the same object! We will observe it, will discuss it, and will let you know. "

Then, the *Master* went to the beach and placed a powerful telescope.

He would not give up until he found out what that mysterious object suspended in mid air was. He remained looking at that object for years, but it was of no avail.

The rest of the inhabitants of the island had already accepted the idea that it was a triangle, as the other naturalists said, and derided both the physicists and the religious who saw different objects. But he was not at all convinced.

A amused little laugh behind him made him turn around.

He blinked to bring the visitor's face into focus and, after a moment, recognized him.

"Ah, it's you!" he sighed, embarrassed.

Much time had passed and that child had become a boy with bright and confident eyes. He was looking at him and laughing while having an ice-cream with delight.

It was the same child that years before pointed out to him the impending object.

"Look, I do not know how to tell you this," snorted the *Master*, "but I have no idea what that object might be. I even asked others, but everyone sees something different. Some people see a circle, others a triangle, others an ellipse with a triangular point on one side..." he explained opening his arms and watching the boy's reaction, who was smiling and licking his ice-cream cone.

"It is a cone" said the boy, raising his cone to the sky.

"I realize it is an ice-cream cone," said the *Master*, "fortunately I can still see that!"

"I am not talking about this cone. I mean, that huge object in the sky... is a cone, like this one!"

The Master smiled with amusement. "You have a very lively imagination, and that is good, but you are wrong: that object in the sky is not an ice-cream cone!"

"You are the one that is wrong, *Master*, because you have no imagination and only reason based on what you can see..."

The *Master* blushed, offended.

"How dare you? I have studied a lot, I am a great academician; you, instead, have an empty mind, just like a white sheet!"

"That is precisely why there is room for new things in my mind," said the boy. "Look, if I put the ice-cream cone like this, I see a circle, but if I turn it sideways it is a triangle, and if I put it obliquely I see an ellipse and a triangular point on one of its sides" he said, grinning.

The Master, stunned, dropped into the sand.

"It was not that difficult" argued the boy. "Why cannot you experts consider everyone's point of view and develop a synthesis?"

The boy was right: the apparent reality could change depending on the point of view, but the essence of the object was immutable:

the object in the sky was a cone, just like an ice-cream cone!

And they could have understood it years before, if everyone had made an effort to listen everyone's opinion.

"And what does that cone represent?"

"It seems clear to me: it is the prejudice, the factious and partial point of view"

"And will we ever be able to get rid of it?"

"Maybe. But it will only melt when heated by the light of tolerance... like an ice-cream cone in the heat of the sun!" replied the boy as he turned and started to walk away.

"Are you leaving?"

"Yes, I am moving to the island of the philosophers. It is much larger than the islands of the scientists and of the religious put together: it is arch-shaped which enables me to have multiple points of view. Moreover, there is a mountain in the center, the *summit of Meditation*, which allows you to see everything from on high".

"Separated as we are from each other, what can we do?"

"You could build bridges that link up your islands, perhaps even to that of the philosophers!"

"It will not be easy" said the naturalist, "I realized long ago that the leader of each island wants to remain a separate island to be in complete control of the inhabitants..."

"Then, try at least to talk occasionally and to exchange your points of view!"

The Master turned up his nose, doubtful.

"But who are you really?" he finally asked. "Tell me who you are!"

The boy made a long pause and munched the rest of the ice-cream cone with delight.

"I am a person with a mind and a heart as white as a blank sheet," he replied with a grin.

Pellegrino De Rosa

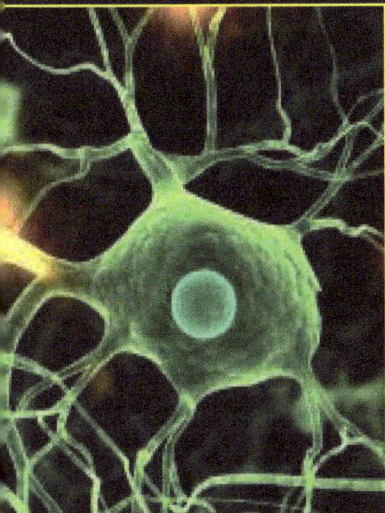

FAQ

Darwinism, Intelligent Design or "Evolutionary Plastcity"?

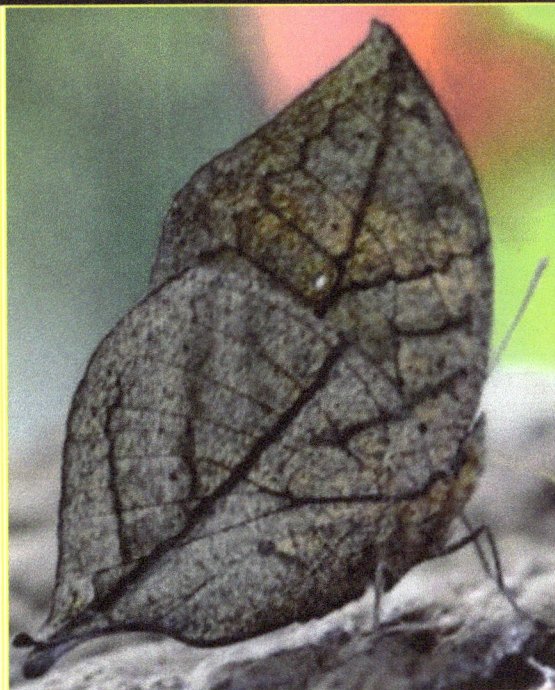

(FAQ gathered by: Benedetta Napolitano, Dante Casoria, Armando Cenni, Francesco Colucci, Antonio De Rosa and Enzo Pecorelli).

What is the "Evolutionary Plasticity"?

The *Evolutionary Plasticity* is a new hypothesis on the evolution of living species presented in 2009 by Dr. Pellegrino De Rosa, Italian writer, journalist and agronomist.

According to this hypothesis, the evolutionary boost would be made up of an alleged *ideoplastic and mutagenic action of the psyche of individuals on the genetic complex* (genome and epigenome) of their reproductive cells.

Is it a creationist or an evolutionary hypothesis?

It is an evolutionary and 'ideoplastic' hypothesis, as hypothesized that evolutionary mutations are caused by the individual's will, but it has some points in common with the classical theories of evolution (Darwinism and Lamarckism) and with the Intelligent Design.

Is it a scientific or a philosophical theory?

It is a theory characterized by a multidisciplinary approach.

It takes detailed ecological and naturalistic observations as a starting point (rapid-mimicry of the *sepiidae*, crypto-mimicry of the *phasmidae*, alleged collective mind of social insects), and then makes some connections with genetics, with cognitive sciences and philosophy.

Then, in an attempt to explain some specific theoretical aspects, it makes reference to some related fields, such as *quantum biology* and *plants neurobiology*.

What is the difference between this theory and Darwinism?

The *Evolutionary Plasticity*, as the neo-Darwinism, accepts the numerous genetic and fossil evidences, and agrees with the fact that living species have evolved and still can evolve, and admits that natural selection leads to the success and survival of the 'fittest' individuals (survival of the fittest).

However, firstly it points out that natural selection only acts once the evolutionary mutation has taken place, and secondly, that the neo-Darwinism does not offer a convincing explanation of how such 'fittest' individuals were formed.

In fact, the *Evolutionary Plasticity* does not accept the blind 'chance' as the source of evolutionary mutations and suggests, instead, that these mutations are directly caused by the *ideoplastic action* of the individuals' psyche on their own genetic material, which not only influences its expression but also produces new genes.

What is the difference between Evolutionary Plasticity and Lamarckism?

Both theories assume that living species have not evolved by *chance*, and that instead evolution was caused by their specific need or desire to adapt.

However, while Lamarckism attributes the evolution solely to a somatic action (*use and nonuse* of parts of the body) of the individual (eg, elongation of the neck of the giraffe), the *Evolutionary Plasticity* attributes it to the ideoplastic action, of an openly psychic nature, which acts through quantum mechanisms, whether undulatory (interfering pattern, collapse of wave function, etc..) or non-local (Bohm and Pribram's holographic paradigms).

Lamarckism does not explain how the somatically acquired characteristics can be transferred to the descendants, while *Evolutionary Plasticity* assumes that the ideoplastic action of the psyche causes the corresponding mutations directly in the genome and in the epigenome of reproductive cells.

Finally, Lamarckism is unable to explain the appearance of evolutionary features not conditioned by the use or nonuse of parts of the body (eg, mimetic coat of the giraffe, shape of the leaf-bug), while the *Evolutionary Plasticity* explains it thoroughly.

What is the relationship between genetics and *Evolutionary Plasticity*?

The *Evolutionary Plasticity* completely admits the effects of genes (and of the epigenetic mechanisms) on the expression of phenotypic traits and the effects of genetic manipulations and therapies on living beings, *but considers that genes are the vehicle instead of the source of evolution.*

In fact, it does not only accept the hypothesis that the evolution of living species is caused by random and uncoordinated gene mutations. And, as it has already been said, it hypothesizes that the psyche of individuals not only conditions the operation of existing genes in an epigenetic way, but can also produce the appearance of entirely new genes.

In addition, Dr. Pellegrino De Rosa suggests that DNA must be studied more and more adopting a quantum and undulatory approach, and suggests that researchers make experiments to verify the presumed mutagenic effects of the psyche on the structure of genes and on the appearance of new genes.

The *Evolutionary Plasticity* is an evolutionary theory that also applies to plants and microbes. Does it mean that they have a mind?

According to the author, it is possible - but needs to be verified - that all the systems with a flow of energy are potentially intelligent (may be even the computers and plasma are intelligent) and since the matter is one of the many states of energy, it may be also be intelligent.

He is neither the first nor the only one that believes so. The author himself has occasionally quoted Max Planck's opinion: "*All matter originates and exists only in virtue of a force... and we must assume there is a conscious and intelligent mind behind this force. This mind is the matrix of all matter*".

Regarding plants, the author makes explicit reference to the observations of Prof. Stefano Mancuso of the University of Florence, and to his line of research on the alleged intelligence of plants (the so-called *plants neurobiology*).

Furthermore, he assumes that it was precisely the mistaken belief that plants did not have intelligence what led - wrongly - Darwinists to consider casuality as the explanation for the evolutionary boost; but also led Lamarckiststo consider the physical principle of *use and nonuse*, creationists to consider the direct action of God, and new-age scholars to consider the presumed and improbable automatic interactions of different kinds as the explanation for the evolutionary boost.

However, the *Evolutionary Plasticity*, taking hypnology as a starting point, considers the *ideoplastic will* of the individual as the main cause of the evolutionary process.

Finally, the author believes that microbes, like the rest of the living beings, were endowed with a mind: which is proved by the simple fact that they manage to survive, to reproduce, to feed and interact with the environment, since the dawn of life on Earth, and still do it with great success and do not worry about the opposite opinions of scientists.

What is the link between *Evolutionary Plasticity* and mimicry?

It is a vital connection. In fact, the theory of *Evolutionary Plasticity* was developed by the author following his own observations on mimetic organisms, especially on the leaf-bug and the cuttlefish.

According to Dr. Pellegrino De Rosa, the rapid-mimetic action of the cuttlefish (which instantly adapts to the color of the ocean floor) has the same psychic and ideoplastic origins of the crypto-mimetic action of the leaf-bug (which has permanently taken the shape and color of the leaves in its habitat). The only difference is that in the second case, individuals managed to fix the adaptation in the genes of their own germinal cells. In conclusion, the mimetic action, presumedly of quantum nature, would be very similar to the process that leads to evolutionary mutations.

Therefore, evolution is merely '*a will that takes shape*'.

This hypothetical process could explain how an organism adapts to the forms already present in nature, on which the psyche may be inspired; but how could it explain the appearance of new forms and functions?

By accessing the information stored in the physical dimension called by Bohm 'implicate order', by Plato's 'Hyperuranium' or 'World of Ideas' and by Hegel 'spirit' or 'idea out of oneself' (cfr. Phänomenologie des Geistes).

This could explain how the orchid *Ophrys apifera* developed an inferior tepal that is similar in shape, color, size, hair and odour, to the abdomen of a bee to attract them and facilitate pollination, and, in addition to this, could also explain the nature of instincts.

The author also highlights that there are some cosmological theories which state that the Universe is nothing more than a tridimensional projection of a bidimensional matrix (conceptually analogous to Bohm's *implicate order*) placed on the edge of the Universe itself, and wonders how come some concepts are usually taken into consideration, at least as a hypothesis of study, in other theoretical areas, and are not considered valid when dealing with the evolution of living species.

What is the relationship between quantum mechanics and 'Evolutionary Plasticity'?

In addition to the link just mentioned (Bohm's holographic paradigm), the author explicitly refers to the studies of Pribram and supposes that the interface body-mind or mind-gametes is holographic and quantum type (bio-entanglement), and presents evidence that could validate his personal interpretation in both the essay and the novel.

What is the relationship between Evolutionary Plasticity and cognitive sciences?

The author, who is keen on hypnology, of the Erickson school, suggested that when an individual is in a state of need or danger, a psychic state similar to *hypnotic monoideism* is triggered (in which, as it is well known, the individual is able to mutate some features of its own body).

The author hypothesizes that in this particular state of mind, the individual's psyche can not only be able to control some epigenetic mechanisms but also to create real and evolutionary genetic mutations.

What is the relationship between *Evolutionary Plasticity* and Intelligent Design?

Both hypotheses refuse that the evolution is determined by the blind chance (and both, among other things, refer to the Haldane's Dilemma).

However, while the Intelligent Design suggests a direct divine intervention of a God, the *Evolutionary Plasticity* suggests a direct involvement of individuals' will, which acts directly on germinal cells using undulatory and holographic mechanisms.

In support of the *ideoplastic hypothesis of evolution*, Evolutionary Plasticity provides some evidence (presumed memory of the organs of transplant patients, hypnotic monoideism, alleged collective mind of insects, striking resemblance of the leaf-bug to the leaves in its habitat, etc..) and states that the mechanisms of action involved are of quantum nature.

The author also emphasizes, in accordance with Giordano Bruno's pantheistic monism, that if God resides in all things - plants, animals and men –, then stating that evolution is produced by the will of the individual, which is nothing but a particle of God, is equivalent to saying that evolution is produced by God.

What would the author say to those who want to place the Evolutionary Plasticity within a psychic neo-Lamarckian category?

The author does not conceal neither his personal liking to Lamarck nor his doubts about Darwin (who, apparently, behaved in an unedifying way snatching the concept of "natural selection" from Wallace). However, he stresses that *Evolutionary Plasticity* is an ideoplastic hypothesis of evolution, and even has points in common with Darwinism and with Lamarckism, it is a separate theory, conceptually different and more complex.

In fact, it is true that it shares with Lamarckism the idea that evolution is not produced by random mutations, but by necessity; and even though redefined, shares with Darwinism the concept of natural selection. But it differs substantially from Darwinism, which concept of random evolutionary mutations categorically refuses, and also differs from Lamarkism in the nature of the evolutionary boost (which according to Lamarck is physical and according to the *Evolutionary Plasticity* is psychic and quantum and acts directly on the DNA).

In addition, the *Evolutionary Plasticity* has a number of fundamental connections with the mimetic phenomenon, with the hypnotic monoideism, with quantum physics and with plants neurobiology - aspects that are present neither in Lamarckism nor in the other evolutionary theories.

After all, do we really need this new theory?

The author often responded to this question in the following way: "We do because in spite of the existence of the leaf-bug, the other theories do not explain thoroughly how the leaf-bug is able to adopt that extremely peculiar shape so precise even in the slightest details. And why should the mechanism that confered it this particular shape be the same mechanism on which the evolution of all living beings is based".

In fact, according to the author, this shape can not be determined neither by lamarckist principles (*use and nonuse*) nor by Darwinist or neo-Darwinist principles (*random mutations*).

He also argues that - while the effects of natural selection are indisputable – it has never been actually proved that random mutations of DNA can lead to complex, integrated and functional mutations. And he emphasizes that even when hybrid or chimeric organisms have been obtained in the laboratory, these variations are not random; they are created by the man.

Therefore, according to the author, the defense of Darwinian positions (regarding 'chance') would seem heavily ideological and lacking any scientific basis, due to the fact that explaining the appearance of complex and functional characteristics by means of a *random event* is a 'non explanation', really unscientific and, above all, statistically unsustainable.

There are hypotheses which state that evolution of living beings may have been influenced by interventions of genetic engineering carried out by extraterrestrials or by crosses between human beings and aliens, what does the author think about it?

The author believes that, at present, this must be considered a science fiction hypothesis, and states that it has nothing to do with the theory of 'Evolutionary Plasticity' (which, among other things, refers to the evolution of all living species and not just to that of human beings).

However, according to the author, it is methodologically incorrect to totally discard this possibility because there is certainly life in other parts of the Universe; and also referring to ancient scriptures, we can not exclude the possibility that our planet has been visited by extraterrestrial beings that can artificially manipulate genes just like we do.

Besides, doubts can be fed by the fact that, as said by geneticists Lynn Jorde and Henry Harpending from Utah University, genetic variation of modern man is very low compared to other species, which proves that the man descends from a few thousands of couples of parents.

Finally, he stresses that any episode of genetic manipulation (alien or human) has nothing to do with the more general process of evolution of species and with the basic mechanisms *Evolutionary Plasticity* has been dealing with.

Considerations on holistic theories.

The author, especially in the novel, seems to believe that all the organisms are 'entangled' (that is to say non-locally connected) between them, and that all theories, including the most fancy ones, deserve to be taken into consideration, at least as a hypothesis of study, because often they are more convincing than those based on chance that, despite everything, claim to be scientific.

However, he stresses that the *Evolutionary Plasticity* is not based on immaterial assumptions, so to speak. On the contrary, it starts by observing living mimetic organisms that are real, not the result of someone's imagination, and that can be observed in nature. And then, since the other theories were unable to provide a comprehensive explanation of their existence, he sought a number of possible explanations that go from *quantum physics* to *plastic monoideism*.

Of course, these hypotheses still have to be proved.

The important thing is to respond the initial question (How was the leaf-bug formed?) and to keep the door open to any possible sign, including those suggested by the most unlikely theories, like that of Backster's primary perception which was reintroduced by some authors with different terminologies and shades (alleged fields of different types, critical mass, familiar constellations, psycogenealogy, internal force, etc..).

None of them is less credible than the one that tries to explain both mimetism and evolution appealing to chance!

This *anglerfish* modified the first ray of its spine and transformed it into a structure called *illicium*, similar to a fishing rod with an worm-shaped fleshy. This evolution can not be attributed to *casuality* or to *use and nonuse* either, but a will that takes shape (*ideoplastic evolution*).

Author's notes

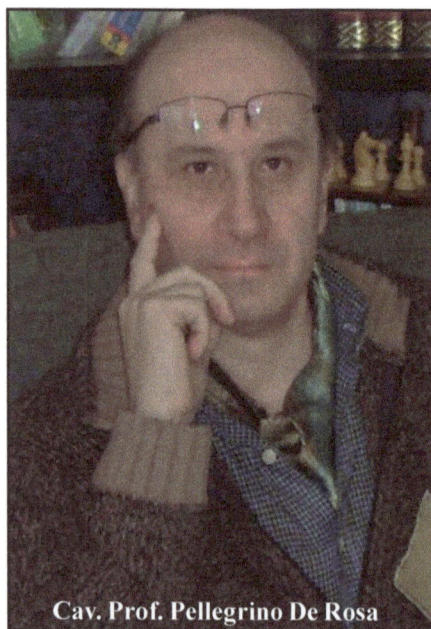

Cav. Prof. Pellegrino De Rosa

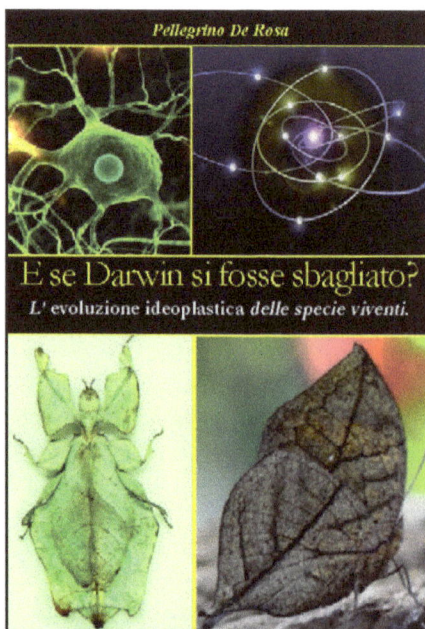

Pellegrino De Rosa

E se Darwin si fosse sbagliato?
L' evoluzione ideoplastica *delle specie viventi.*

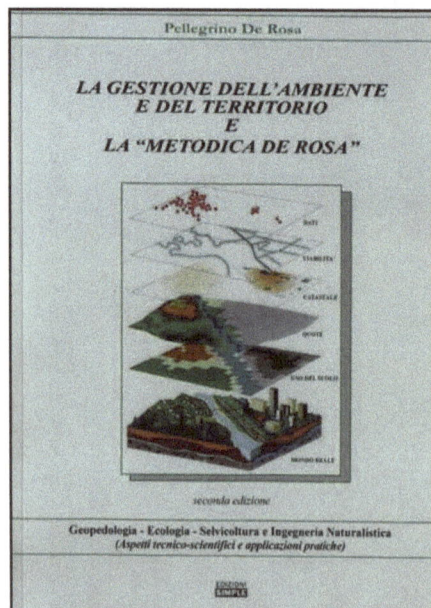

Pellegrino De Rosa

LA GESTIONE DELL'AMBIENTE
E DEL TERRITORIO
E
LA "METODICA DE ROSA"

seconda edizione

Geopedologia - Ecologia - Selvicoltura e Ingegneria Naturalistica
(Aspetti tecnico-scientifici e applicazioni pratiche)

EDIZIONI
SIMPLE

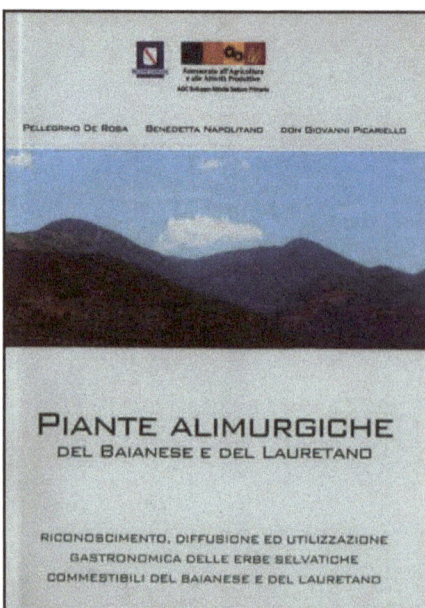

PELLEGRINO DE ROSA   BENEDETTA NAPOLITANO   DON GIOVANNI PICARIELLO

PIANTE ALIMURGICHE
DEL BAIANESE E DEL LAURETANO

RICONOSCIMENTO, DIFFUSIONE ED UTILIZZAZIONE
GASTRONOMICA DELLE ERBE SELVATICHE
COMMESTIBILI DEL BAIANESE E DEL LAURETANO

About the author

Pellegrino De Rosa is an agronomist, journalist, essayist and writer. He graduated in Agricultural Science and Technology, and has a master's degree in Management and homeland defense. He has a multidisciplinary background: run some local newspapers and has been interested in computer science and electronics.

He is an agronomist and teaches science and technology at secondary schools.

He is interested in design and in environmental, botanic, faunal and naturalistic engineering studies. In this latter sector he presented a particular methodology suitable for the Vesuvian pyroclastic soils.

He has published several ethnographic and technical essays, including a text on *alimurgic plants* (edible wild herbs) published by the Campania Region. He is interested in the Ericksonian hypnology and is an instructor of elementary chess.

He was awared the high honor of 'Cavaliere al Merito' of the Italian Republic.

To contact him:

Email: studio.derosa@alice.it

Facebook: Pellegrino (Rino) De Rosa

Other publications of the author

Alimurgic plants (Botanic and etnographic study on 74 species of edible wild herbs and their culinary use). Printed by the Campania Region - Department of Agriculture and Productive Activities. Prestigious preface by Prof. Antonio Saracino of the University Federico-II, School of Forestry and Environmental Sciences of Portici (Naples).

Environment and Land Management and the "De Rosa Method". Environmental study on the inland areas at risk of hydrogeological instability.

Presentation of an innovative methodology on Naturalistic Engineering to be applied in pyroclastic soils and in those presenting inert pedologic horizons due to physical, chemical and biological causes. (*ISI Web of Science - Thomson Reuters. Patent Number IT1372222-B, International Paten Classification A01G-000/00, 14 Jun 2006*).

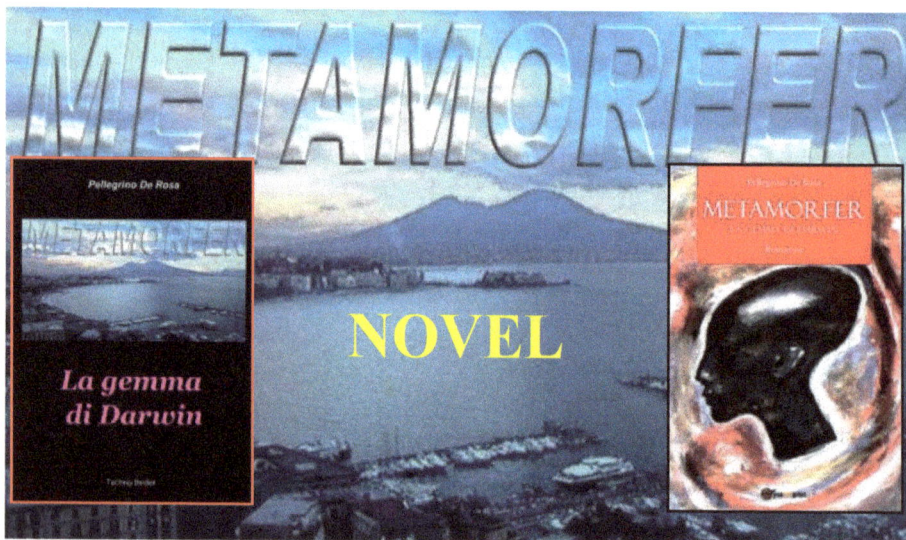

Metamorfer. The gem of Darwin. (Novel)

(Review by Emanuele Pettener, writer, essayist and professor of Italian Language and Literature at the Florida Atlantic University of Boca Raton- Florida- U.S.A)

Gulf of Naples
Fresh air, rough sea, sensual atmosphere.
Appears a grim and fascinating character, Raf, longing for vengeance.
Now a beautiful reporter appears, Eva Nabokova, her hair is the color of ripened wheat.
And then a number of mysterious questions: Who or what struck the Newfoundland dog? Who shot the dolphin? Who blew up the nomad camp at Ponticelli? And who is the mysterious creature Raf was trying to release taking the chip of the "gem of Darwin" with him?
Pellegrino De Rosa's debut novel does not make you waste time: it keeps you glued to it right from the first lines and keeps you walking a tightrope at the brisk and exciting pace of the best action-movies till the last and unexpected chapter.
The main story is entangled with many other stories: that of a cute press photographer, inveterate womanizer; that of a sexy Italian-American spy; that of a mysterious old Neapolitan gypsy; that of a group of "femminielli"

(folk transvestites that prostitute themselves) and a nostalgic boss of the Camorra, and of many other characters, mostly supporting, but they all were carefully delineated and had pathos.

And the magic Naples in the background with its alleys, its smells, its legends and its extremely colorful characters.

The most important thing, a new evolutionary hypothesis is presented (the Evolutionary Plasticity) which correlates evolution, mimicry and quantum sciences.

But, the complexity of the argument does not make the novel dull reading at all; it indeed flows lightly and cheerfully like clear water on a mountain stream.

In fact, the author managed to combine - with a pleasant style and supreme lightness - landscape paintings, Neapolitan witticisms, folk legends and eroticism with action, mystery, science and philosophy.

The end result is an Italian fantasy-thriller that, as for content, suspense and humor, can decently compete with foreign giants of the same genre, and with a cut above: the unconventional and fatalistic Neapolitan irony.

(Review by Albino Albano, professional journalist, correspondent of the newspaper "Corriere" of Irpinia).

This is a novel of 382 pages full of action, irony, colors and landscapes; it is very easy to read and attractive. Then, by page 150, the story takes off permanently and the pace fastens, shrouding readers in a fun and engaging atmosphere reminiscent of Fleming's spy-stories.

And then one realizes this is a real masterpiece.

The description of Naples, alive and vibrant; the complex and yet fluid argument, like a crosscutting film-script, and the characteristic and charming characters make this novel ready for a film adaptation.

The description of the Golden Mile, the stretch of the Vesuvian coast between Ercolano and Torre del Greco is remarkable. The description of the Gulf of Naples and its beautiful grounds is really magic and sparkling. Multisensory and evocative is the description of the mysterious caves and the packed alleys of Naples with its sounds, its smells and its people so nice and positive.

The accurate and detailed description of the archaeological places, of the scientific equipment and the naturalistic implications is also impressive.

This novel has several layers of meaning; It is nice and profound at the same time. I recommend you to read it.

The novel is available in ebook and in the best bookstores on-line.

youcanprint

Finito di stampare nel mese di Giugno 2015
per conto di Youcanprint *Self - Publishing*